Contents

A Future With a Silver-lining

Silver Magic

How Colloidal Silver Can TRANSFORM Your Life

By C.K. Murray

Join the Newsletter

Similar works by C.K. Murray:

Master Mind: Unleashing the Infinite Power of the Latent Brain

Coconut Oil Cooking - 30 Delicious and Easy Coconut Oil Recipes Proven to Increase Weight Loss and Improve Overall Health

-

Health Hacks - 46 Hacks to Improve Your Mood, Boost Your Performance, and Guarantee a Longer, Healthier, More Vibrant Life

The Omega Factor - 20 SUPERCHARGED Omega-3 Recipes for the Body and Mind

Body Language Explained: How to Master the Power of the Unconscious

Natural Weight Loss: PROVEN Strategies for Healthy Weight Loss & Accelerated Metabolism

Neuro-Linguistic Programming Explained: Your Definitive Guide to NLP Mastery

They've tried to keep it from you.

For decades, the Medical Orthodoxy has denied it. They've decried it. It doesn't work, they say. It's dangerous, they warn. It's just more "quack" pseudoscience, they claim. They've called believers of it fools. They've called sellers of it "snake oil salesman." The FDA has closed the book on it, because *obviously* it's "not generally recognized as safe and effective."

And this is the plan. Has been for some time. Everywhere you go, they've hushed, ignored, and undermined any and all people who dared to believe, to know, that it is good.

But now it's too late. Because finally, the truth is unleashed.

"Silver Magic," aka Colloidal Silver, is here to stay.

And always has been! The problem is, many people have been *brainwashed.*

In an effort to keep this amazing product out of your hands, Big Pharma and other Power Brokers in the Medical

Orthodoxy have force-fed us pills, drugs, and all sorts of risky, potentially deadly chemical compounds that not only kill the 'bad stuff,' but potentially the patient too.

You know them, you've seen them. Chances are, you've taken quite a few of them yourself.

They are disseminated based on dubious clinical studies, regurgitated through T.V. talking points, and pushed upon physicians and specialists everywhere for one, primary reason: $$$

You've seen the glowing commercials with the happy faces and syrupy voices. A prescription for all, a solution for none.

'If you have *this*, take *that*.'

'Ask your doctor about *this*.'

'Tell your doctor about *that*.'

'If X doesn't work, take Y.'

'And if Y doesn't work, consider Z.'

Or maybe you should just be on X, Y, and Z all at once. Why not, Yea?

And Heaven forbid, if *X* or *Y* or *Z* happen to cause distress, happen to give you cancer 20 years down the road, or sudden unpredictable heart problems 5 days later, or some other acute adverse reaction that won't go away, or heck, even (in rare cases) result in premature death, then, well...

T R U S T   U S

That is all they want. For you to trust them. To buy into their system, unquestioningly. If you get hooked, addicted, *torn apart* by their powerful pharmaceuticals, so be it. This is merely the risk. Succumb to their ways, and mindlessly obey. Accept physical malaise, accept cerebral fog. Because slowly, but surely, they're bringing you down...

But you can rise. You can stand. So stand and think, open your mind, shift your schemata and take action. Never, ever buy what they're selling at face value. Look deeper and think clearer. There is an alternative to their ways.

And they know it. So seize it, take your chance and don't delay.

It's time to change the game, it's time to transform your life. It's time... to experience Magic...

***

Why the Bad Rap? The Historical Truth of Colloidal Silver

The first question you might be asking is: *why*? If colloidal silver is so important, you ask, so all-amazing, *why* have the powers-that-be hidden it from public knowledge? Why do they call it dangerous or ignore it altogether? Why do they bury studies about its incredible broad-range health benefits? Why don't they market it and capitalize on its powers? Why not harness silver for what it is? - the single most powerful, natural infection-fighter in the world?

Well, you just answered it! The Medical Corporatocracy doesn't *want* you to have that power. They want you on mediocre 'treatments,' all sorts of pharmacologic prescriptions that work, but not well, that oftentimes require other pharmacologic treatments just to counteract *this* side-effect, or modulate *that* adverse reaction.

If silver is as good as the demonized research indicates, then *of course* the Medical Power Brokers don't want it getting out! Why would they sell you a product that takes care of your problems with hyper effectiveness? What would be the point in that? The best way, the Number One way, to monetize allopathic medicine is through *continuous drug regimens*. The Medical Orthodoxy knows that if it can get you on a *continuous drug regimen*, it can slowly drain you. Instead of actually addressing the core source of your health maladies, the Medical Orthodoxy merely treats the symptoms. They can leech you for much more money, keep you in a dependent, semi-healthy state, and maximize the rate of side-effects – that are then treated by more of their drugs.

In many ways, their drugs aren't necessarily improving your health in the long-run, they're just providing you a temporary relief. Because then, the antibiotic you were using becomes ineffective, the bacteria becomes resistant, and your only choice is to pump yourself full of *another* pharmaceutical with *another* host of side-effects, until *another* period of time passes and you need *another* drug to get… 'healthy.'

But that's the plan, that's the game. Because in the end, it's not so much about making you better, as it is about *masking* the fact that you're not getting better.

Treat the Symptoms. Ignore the Source. ~ The Medical Orthodoxy Mantra

Here are the facts: If you read and *believe* the mainstream stuff, you'll find that all silver products are potentially lethal. That your skin will quickly turn blue and grey—argyria—and you'll soon be one metal piece away from the tin man.

If you read and believe the Medical Orthodoxy Narrative, you'll 'know' one thing. That there are no discernible health benefits whatsoever to silver, and that anyone telling you that there are, is selling you a line of bull.

False.

Truth be told, silver, in its many forms, has been used for hundreds and even thousands of years. Its powers were known to many of the ancient civilizations we study today. The Greeks harnessed and cherished the preservative effects of silver. They stored water and foodstuff in it, and Hippocrates, one of the greatest physicians in human history, was the first to note silver's antimicrobial properties in 400 BC. He spread word of silver's greatness, claiming that it could heal most wounds and tame most diseases. Even the king of Persia, confident in silver's abilities, traveled at all times with water in silver vessels to prevent sickness.

The Romans were no different. The upper echelons relished silver in both vessel and utensil. The Egyptians loved silver so much they wrote about it over and over! Or how about the Chinese emperors, dining with silver chopsticks? Or the Middle Ages, when silver utensils and goblets and hedonistic silver consumption gave the upper class a bluish tinge to their flesh, thus resulting in the term "bluebloods"?

But the cost of blue skin was well compensated. These bluebloods, born with a silver spoon one might say, were *noticeably* resistant to the deadly plagues that ravaged much of Europe at the time.

Even coming into modern times, the Imperial Russian troops of WW1 used casks of silver to sterilize water from local bodies of water. This use even extended to WW2, wherein forces of the Soviet Army employed silver in its many forms for preservation, meals, and medical treatment. All across the world, and the span of human history, silver has been harnessed for good. The Ayurdevic healers of India, for instance, still use silver today to heal both mind and body.

Silver in the Western World regained momentum into the early 1900s and heading in the years prior to WW2, the use of silver as a powerful antimicrobial and antifungal spread far and wide. At one point by the mid 1940s, there were roughly 50 different silver compounds being applied in topical, oral and even intravenous and subcutaneous forms. Despite several mishaps due to argyria caused by unsafe, contaminated and inferior silver compounds, silver as a sort of cure-all became prevalent.

Unfortunately, antibiotics would quickly end this rise to the top. Silver fervor dissipated quickly, as manufactured synthetics targeted microorganisms with supposed ease. By the third and fourth decade of the 1900s, antibiotics were on the rise, with today's resistant 'superbugs' still a thing imagined.

However, the pendulum would once more swing back. As modern researchers studied new effective treatments for physical ailments such as burns, sores, wounds, and infections, silver again returned to the fore. And as once-unforeseen superbugs became reality, the 'all-powerful' antibiotics began to lose their luster…

By the later decades of the 1900s, silver was reasserting itself as a powerful and effective treatment for all. Continued research revealed that silver was a destructive force against numerous strains of bacteria, viruses, and fungi. Yet, at the same time, silver appeared to be non-toxic to the human body, even as the many compounds and solutions of silver destroyed all else. As the modern day approached, experts and laymen alike were learning: silver, in its many forms and functions, was a magic all its own.

Nowadays, this should be no surprise. Silver has been, and always will be, a critical factor in the health and well-being of the human condition. As far back as records exist, and silver was known, minds have studied and applied the metal's benefits. It is not BAD, it is GOOD. It is better than GOOD, it is (in some cases) MIRACULOUS.

So when the Medical Orthodoxy tells you that silver is not what it claims to be, or that it is ineffective, or that it only works at times, under certain conditions, and thus is not worth the time – they are LYING. They will spin, distort, and obfuscate the truth to keep you dependent on what *they* produce, and *they* sell, and *they* profit from, time and time again.

So don't be duped. Don't be pressured. Don't be rendered inferior.

Just know this: silver is, and will continue to be, a magic all its own.

However, let's not be hasty. Because magic is not necessarily medicine. At least according to the Medical Orthodoxy. See, in 1999 the Food & Drug Administration (FDA) issued a ruling into law that rendered any silver substance in a gelatinous solution unable to be marketed, branded or sold as a medicinal substance. Colloidal silver was categorized as an unclassified drug, and any colloidal silver products claiming to be of medicinal benefit were directly violating the federal law.

The only good news was that silver could still be protected, thanks to a pertinent act from 1994. Still applicable today, the Dietary Supplement Health and Education Act (DSHEA) provides that silver be labeled as a dietary supplement, due to its status as a natural substance, not a controlled substance. In 2002, the FDA issued revised guidelines allowing dietary supplements to be sold with specific health claims, given sufficient scientific evidence for such claims. The "Guidance for Industry: Qualified Health Claims in the Labeling of Conventional Foods and Dietary Supplements" signaled that allowance.

So here we are today. Surrounded by potential for amazing things, powered by a force in silver that is almost incomprehensible in its effects and versatility.

Many experts have even contended that colloidal silver is so powerful against bad microbes and fungi, that no microorganism can survive direct exposure beyond 6 minutes. It is especially powerful against bacteria. Research indicates that the main reason colloidal silver is so powerful against bacteria is because, unlike antibiotics, colloidal silver does *not* target the bacteria head-on. Instead, silver attacks the *food source* of bacteria, essentially starving them until they are rendered inactive.

It is an amazing find, but one that may help explain why colloidal silver is so beautifully reactive when in the presence of vicious microbes.

To put it simply, colloidal silver doesn't play games. It is fierce. It is smart. It is strategic. And it is natural. And it can be used in a number of fashions, a number of forms, a number of ways, to heal seemingly everything under the sun.

And heal, *heal, h e a l* it does!

***

What is Colloidal Silver, Anyway? Distinguishing Silver from Colloidal Silver

We've all heard of gold and silver. They are the top two medals at the Olympics. They are shiny, glossy, heavier elements. They are beautiful and hard and signal wealth and success. They have a history, they have an aura, they represent something different to everyone, but something similar to everyone too. Gold and Silver are amazing.

But while Silver may take runner-up to Gold in contests, its powers for health and well-being are second to none.

First off, let's define silver for the… 'unenlightened.'

Silver is a white lustrous metallic element found all over our natural environments. Low concentrations of silver can be found all throughout the human body due to inhalation of air particles, and the ingestion of food and drinking water with trace metals. Silver, in its many forms, has been increasingly applied for therapeutic, antibacterial, antiviral and antifungal uses, whether for wound care dressings, in medical devices such as catheters, surgical sutures, protheses and dental fillings, in textiles, in cosmetics, and even throughout domestic appliances.

Put simply, silver – in varying amounts – is everywhere.

However, silver and *colloidal* silver are two different things entirely.

While silver can be tweaked and manipulated, combined and collated, in countless different ways, one of its finest forms, and most powerful and effective, is that of *colloidal* silver.

Colloidal silver is a solution of water containing microscopic particles called nanoparticles. The silver is suspended in the water due to the small size of the particles, and the total volume of silver is expressed as milligrams of silver per liter (mg/L) of water, or parts per million (ppm)

What's important to note is that the total silver content includes two forms of silver: silver particles and silver ions. Superior colloidal silver products use a balance that is usually 80% silver particles and 20% silver ions, containing no additives or artificial ingredients. Silver particles are tiny spherical pieces of metallic silver, whereas silver ions are atoms of silver lacking one electron. This removal of one electron alters the physical properties of the atom *significantly*. Silver particles will never dissolve in water;

silver ions *need* water to exist. They cannot exist without being dissolved in water.

But what matters is this: colloidal silver is mostly silver *particles*, not ions. It's the particles you want and need, and it's the particles that will imbue you with their unbelievable powers.

Of course, the restorative, healing, preventative, empowering, strengthening and immunomodulatory powers of silver can always change. That is, depending upon how *you* change.

See, the Medical Orthodoxy will tell you that silver is dangerous because if left unregulated it can do damage. Well geez, if you take too much Aspirin you'll do damage! If you chew too much gum you'll do damage. If you drink too much infused water you can literally drown in your own body from hyponatremia! Everything in moderation is the key, and with silver, you sure as heck won't need much.

But first… let's learn what you *do* need, and in what way you need it.

When it comes to bioavailable forms of colloidal silver, there are three main types that supercharge the body and mind. Remember, true and real colloidal silver is mostly microscopic particles—nanoparticles—of silver in water or some liquid medium. Because silver naturally occurs in food and water, this is the best delivery mechanism for introducing silver into the body.

And there are a number of ways silver can enter the body. So before we discuss the most effective, life-altering colloidal silvers known to man, let's discuss how this fancy-schmancy silver thing gets into your body in the first place.

You might be surprised...

***

How Does Colloidal Silver Get Absorbed?

If you want colloidal silver to go straight to your blood stream, there are basically two ways to do it.

One, is to use a nebulizer, which is a machine often used by people inhaling asthma drugs. What the nebulizer does is convert the colloidal silver nanoparticles into larger droplets that are then inhaled. The small droplets then pass through the lung tissue and go straight into your blood stream where they circulate and visit the many organs and corners of your body. Meanwhile, the silver *ions* produce byproducts that are sucked from the blood stream by the kidneys and expelled in your urine. Silver ions do not exist long in the body, no matter what you do, or how you introduce them.

The second delivery mechanism for colloidal silver into your blood is sublingual. Basically, you have to introduce colloidal silver under your tongue, as those membranes are very absorbent. Many forms of colloidal silver are taken under the tongue, whether through a teaspoon or an eye dropper. If you don't want to directly target the sublingual area, you can just swish colloidal silver or nanosilver solutions around your mouth. Inevitably, you'll absorb some and reap the benefits.

Of course, there are also a *number* of other ways to introduce colloidal silver to the body. Even if you don't get it in your blood stream, you can apply it topically to areas on your flesh. You can put it in your ears, your nose, between your toes, and even in your eyes!

And believe it or not, all of these methods, all of these delivery mechanisms for high quality colloidal silver, are not only harmless, but totally, unbelievably *effective*.

Still don't believe it? Worried it's too risky? Still not convinced? Afraid that you might be increasing the chance of silver toxicity?

Well worry no longer. It's time to debunk some of the common myths (and scare tactics) perpetuated by the Mainstream Medical Orthodoxy, funded by Big Pharma, and echoed by Power Brokers.

***

But Can It Be Toxic? Debunking Mainstream Silver Myths

You've been told that it could kill you. You've been told that it isn't good for you. You've been told that even if it doesn't kill you, and even if it isn't necessarily bad for you, it still will do nothing for you. You've been told a lot of things, and almost all of these things – if not all of these things – are false.

In a comprehensive U.S. Public Health Service report from the Agency for Toxic Substances and Disease Registry, the toxicological effects of silver are designated at a negligible level. It is reported that silver that is eaten or inhaled exits the body through feces in roughly a week. Researchers even report that argyria, the blue-grey discoloration of the skin caused by too much silver, does not pose any real health risks. Simply, it's a cosmetic issue, and one that only occurs if you're ingesting inhumanly large amounts of silver that is contaminated and inferior. And most experts agree that the risk of developing argyria is so low, it's basically fear-mongering to suggest it will happen.

And at the end of the day, scientists say that argyria is the most severe reaction one could conceivably have. Other researchers agree, stating that silver exhibits very low toxicity, whether through inhalation, ingestion, dermal application, urological or blood stream delivery mechanisms.

One prominent study directly measured the effects of administered silver nanoparticles on human subjects. The participants underwent a battery of tests for metabolism, blood profiles, urine, lungs, chest and abdominal MRIs. In the end, the researchers contended that no clinically significant changes occurred *in the metabolic, hematologic, urinary, physical or imaging findings. N*o toxicity at all was exhibited anywhere in the human body during the 14-day trial.

Amazing!

But wait, what does this all mean? Well, to put it simply, it means that researchers are aggressively *trying* to find terrible things wrong with human silver use – but they can't!

Even the U.S. Public Health Service can't find anything serious, and they've extensively reviewed and studied the most current, innovative research inquiries into the manner. It's an amazing thing, that something like this could be so specifically targeted at the bad microorganisms in and

around your body, and be so effective at destroying them –
while also avoiding the *good* microorganisms all inside and
around you too!

But of course, not all forms of silver are effective. The more
you look into silver, the more you find there are so many
combinations and permutations and manipulations that your
head will spin!

So stay steady. Stick to colloidal silver – *the* leader in
powerful silver – and use it like your life depends on it.
Because in many ways, it does. Colloidal silver has the
potential to totally transform and supercharge your daily
living. It has the power to utterly revamp your body's
immunological responses to everyday microbes, to
infrequent superbugs, and in the face of devastating
epidemics.

But first, you've gotta know your stuff. If you want to enjoy
all these unparalleled benefits, you need to get the right
silver.

Don't delay. It's time to learn what, exactly, makes *superior*
colloidal silver magic, and what makes all the other colloidal
silvers, mere pretenders…

***

Forms of Colloidal Silver - Find the Best, Forget the Rest

Don't fall for meaningless pseudo-babble and gobbledygook. Don't be swayed by cunningly slick and deceptive marketing schemes. And never, ever, succumb to the power of an *inferior* product dressed up with superior-*looking* labels.

It's simple. Not all "colloidal silver" is colloidal silver. Many manufacturers will fudge their claims. False advertising will tout 'natural' high-powered colloidal silver, when in fact you're receiving a product of low-cost, low-standard manufacturing that is mediocre at best.

Don't be swindled. Here's what you need to know:

Ionic Silver

First thing you need to know: this is not colloidal silver. Many will market is as such, but it is not. Fact is, silver ions and silver particles are different. And they are different for very important reasons. As previously stated, a silver ion is an atom of silver lacking one electron. A silver atom missing an electron is a silver ion. And these are not, not, not the same as silver *particles* that comprise true, high-quality colloidal silver. The important thing to note is that colloidal silver features nanoparticles of metallic silver, and these particles are independent. They do not combine with other elements. Ionic silver, by contrast, does, and it does so readily.

So why is this an issue?

The reason this is bad is because ionic silver will quickly and easily react with other elements to form new compounds. When you take in silver ions, and not silver particles, you are essentially introducing a worthless silver into your body. Why? Because these silver ions will quickly combine with chloride to form silver chloride. This compound is almost immediately eliminated and evacuated through urine. Thus, you get little to no benefit and basically paid for a product for nothing.

Don't pay for nothing.

Typically these products are only 10% silver particles with the rest insoluble silver ions. Now this isn't to say that ionic silver isn't useful. It has demonstrated power against microbes, but much more when outside the human body, not in it. So if you're looking for a good sanitizing to clean your place of residence, ionic silver will work perfectly. But if you're looking for a game-changer for your health and wellness...

***

DIY Lab – Analyzing Ionic Silver

But look, first you have to know exactly what you're dealing with. And if you're still not sure what kind of silver you're dealing with, test it. Grab some table salt and throw in a few shakes. If the solution you purchased turns cloudy white, you're dealing with silver ions, not silver particles – not good!

Of course, not even this always works. In rare (but unfortunate) cases, the solution you have won't turn cloudy white. Why? Because the solution contains no silver at all! That's right, some manufacturers will totally beguile consumers by selling a product that has no silver, as colloidal silver. Talk about predatory marketing…

Which is why you need to know even more about colloidal silver before making an informed decision. Here's the next type of so-called colloidal silver:

Silver + Protein Binder

These so-called colloidal silver products are actually a combination of silver particles and protein binders. Although many of these products will be labeled as colloidal silver (with no mention of protein) they are actually far from colloidal silver. The problem with silver protein products is manifold.

Firstly, these products contain larger-than-normal particles of silver, which is why protein binders are required to keep the particles from sinking in water. Many manufacturers use gelatin as the protein binder, which is bad because (1) it increases the risk of bacteria, and (2) it has a low absorption rate like silver ion products. Bottom line: silver with protein is an inferior product that is not only largely ineffective, but <u>potentially dangerous</u>.

Another thing to keep your eyes on is the concentration of the product. A lot of manufacturers will brag about how high the silver concentration parts per million (ppm) is, stating that this concentration of all-powerful silver makes the product superior.

Well this is false. Not only false, but a common tactic used to market inferior, ineffective silver products masquerading as high-quality colloidal silver. What you need to know is that silver concentration (ppm) is irrelevant. What matters is the 'particle surface area,' not the silver concentration. Besides, exceedingly high silver concentrations (110+ ppm) can, and do, lead to actual blue-gray discoloration, known as argyria.

So then what is 'particle surface area' you ask? Particle surface area refers to the surface of each and every silver particle in the product. It is through the surface of each particle, that the silver interacts with its environment. The silver on the inside of the particle, however, does not react with the environment.

In other words, when a particle is very large, all of the silver inside that particle is wasted. This is why very small nanoparticles are preferred. Not only do very small particles require a much smaller concentration for effectiveness, but they also present a much greater surface area. If any of this is confusing, just remember, in most cases: high concentration = low surface area; low surface area = ineffective product.

Effective superior silver products may have much lower concentrations, but because the particle sizes are so much smaller, you get a lot more 'bang for your buck.' Thus, there is no point in having large silver particles because not only do they (a) require proteins susceptible to bacteria, but they also (b) require exceedingly high concentrations which can lead to argyria.

DIY Lab – Analyzing Silver + Protein Binder

So be mindful. If you think you have a silver protein product on your hands, and not true colloidal silver, just look. There are a number of simple visual tests you can conduct to detect an inferior predict. First, check the concentration (ppm). Silver protein products often contain more than 500 ppm. Next, check the color. The substance will look anywhere from amber to dark grey or black. Finally, give the substance a hearty shake. If it foams up, it contains a protein binder.

And there you have it.

Alright, now let's move onto the final type of so-called colloidal silver. Fortunately, this one is the real deal. It's the bee's knees. It's the optimal silver product for human health and well-being. That's right, I'm talkin… 'Magic.'

***

PURE Silver Magic

In the marketplace for silver, there are a lot of pretenders. And many of them will use pseudo-scientific terms and labels to deceive you into purchasing their product. Truth is, if you want Silver Magic, real colloidal silver, you're going to have to look. Take the time, do the searching. Superior colloidal silver is the product of a focused, complicated and high-cost manufacturing process. It is the real deal. And frankly, there is *nothing* else like it.

Superior colloidal silver contains more than half silver particles. Sometimes as much as 80% particles, 20% ions. There are no protein binders, and it is in high quality deionized water, not gelatin. This is critical because pure deionized water has had virtually all mineral ions removed. This means there are no aspects of sodium, iron, calcium, copper or anything else to potentially modulate or contaminate the superior nanosilver suspended therein. These powers of optimized H2O are indispensable to ensuring superior colloidal silver.

Superior colloidal silver also has a very high particle surface area. It does not, however, have a very high concentration (ppm). The nanoparticles are small, very small, and they are waiting to work magic on, and in, your needing body. There are no additives in superior colloidal silver.

***

DIY Lab – Analyzing PURE Silver Magic

Be mindful of two things when looking for superior colloidal silver. (1) Note the concentration, as optimal colloidal silvers typically range from 10 – 30 ppm, contained in purified water. (2) Check the color. Silver particles block light, so superior colloidal silver will not be crystal clear but darker. If all of the aforementioned attributes check out, you, dear reader, *might* have high-quality colloidal silver on your hands.

But there is more to learn.

Because at the end of the day it's all about quality. The quality of colloidal silver is about the efficacy of the product, how well it works, the efficiency of the product, how much is needed for a given benefit, and the safety of the product, how likely you are to have negative effects. Now, for superior colloidal silver, you won't have to worry too much about safety issues. You can't get argyria from true colloidal silver, and unless you're guzzling the stuff like the Silver Surfer, you're going to be fine. In fact, with superior colloidal silvers (10 – 30 ppm) you can easily take half a teaspoon a day for health and well-being maintenance. When sick or feeling down, you can take a full teaspoon daily without worry.

But before you worry about how much to take, or when to take it, you've got to understand the product directions. And before you can understand the product directions, you've got to find the brand for you…

***

Smart Consumption – Finding the Optimal Silver Product for YOU

When searching for TOP colloidal silver products, it can get overwhelming. Every brand claims it has the highest standards, and it is easy to fake glowing reviews to artificially bump certain products up in the algorithms of online stores. Without naming any names, there are plenty of brands that are seemingly 'the best' (based on 5-star reviews and prevalence) but fail to measure up when actually analyzed.

So if you're looking for a product that is truly top quality, and not a bunch of hot air, look no further than Colloidal Science Laboratory, Inc. The experts at this laboratory have carefully analyzed numerous brand and off-brand colloidal silver products, collating the results in easy-to-read tables.

And one of the best of the best, is none other than MesoSilver 20. According to Colloidal Science Laboratory, Inc., MesoSilver 20 has a 100% accuracy of product labeling rating, the highest Particle Surface Area, the highest Efficiency Index, the lowest Quantity of Colloid (mL) required for Particle Surface Area, *and* the lowest Cost per cm² of all tested U.S. colloidal silver products…. In other words, it's not only a superior product, but it comes at a tremendous value! Another good colloidal silver product (but far from MesoSilver) is **Utopia Advanced Col. Sil. 20.**

And if you're looking for products to AVOID, make sure not to touch "Super Silver Solution," "Oxysilver," or "**Health & Herbs Col. Silver 10**" with a ten foot pole.

Alright, so now that we've identified some of the best colloidal silver products and what distinguishes the Magic from the Myth, it's time to take a deeper look. Let's go behind the scenes, and understand exactly what this Magic does…

Like Magic – the COUNTLESS Benefits of Superior Colloidal Silver

Silver has long been a miracle-worker in many ways. The only problem? It wasn't patented. Big Pharma couldn't capitalize on it, and antibiotics were coming into the fray. But antibiotics have an unfortunate side effect that silver does not. For one, the more we use antibiotics, the more resistant bacteria becomes, and the less effective the antibiotic becomes. Besides, antibiotics not only kill 'the bad stuff' but the good bacteria too. Hence the name, anti*biotic* – they go against *anything* that is *biotic*, or living.

Colloidal silver, on the other hand, does not have a demonstrative resistance effect on these organisms. In other words, unlike the often overpriced, uncovered, overly prevalent pharmaceuticals on the market, colloidal silver does not predictably strengthen the bad 'biotics' it's supposed to be fighting. In fact, research shows <u>little to no increased resistance in microbes</u> when using high quality silver products.

But that's not all. In fact, the powers of colloidal silver against bacteria, viruses and fungi are just the beginning. So let's take a look, at the numerous incredible health and everyday benefits of what can only be described as, Silver Magic, the *panacea*…

(A) Eye infections – If you've ever had conjunctivitis (pink eye) or some other kind of eye infection, you know how much it sucks. Your eyes are bleary, bloodshot, your vision poor, itchiness, soreness, aching, pain, pain, and all sorts of nasty tear-filled discharge.

So give up the eye drops that only make your eyes drier in the long-run. Give up the steroidal treatments that only experience diminishing returns. Don't be nervous about using colloidal silver in your eyes. Remember, you're not dropping shards of metal in your eyes. These are healthy, powerful nanoparticles in parts per million. They're great for your eyes. Research even shows that silver in the eye can stave off, and treat, a number of degenerative retinal diseases.

It's easy. Simply put one or two drops in the eye, three to four times a day until noticeable lessening in symptoms.

(B) Sore throat & Strep – We have all experienced this at one point or another. And what do we do? A lot of us turn to cough drops or throat lozenges or other naturopathic aids at that first hint of rough, itchy discomfort. But what does this do?

Sure, sometimes it works. But most of the time it merely lessens the symptoms without treating the cause. And then what happens? Well, if the symptoms don't lessen enough, or the problem doesn't go away, you turn to antibiotics. And while a powerful antibiotic such as 'Zpak' might knock out the bad stuff, it also kills off many of the advantageous bacteria that would help fight off Strep in the first place.

This is why silver is key. A high quality colloidal silver functions as a <u>consistent and virtually risk-free bactericide</u>. It does not contribute to resistance in microbes, and it does not upset the delicate immunological balance of your body.

So use silver. All it takes is a teaspoon, gargle it for roughly 1-2 minutes then swallow. That's right, swallow. Repeat several times a day, and if you're using a spray form or silver, spray at least 5 times an hour. This sounds like a lot, but the dosing is actually small, and it's ultimately much healthier than turning to a needless pharmaceutical (and it will also save you $$!)

(C) Wounds, Soreness & Inflammation – This one is just, plain, cool. See, the restorative powers of silver are no myth. In fact, silver has long been known for its use in infected wounds, especially during wartime. Silver nitrate was used with, and without, dressings throughout the 1800s for burns, ulcerations, and wounds and was documented for a number of incredible benefits. Not only did it reduce the wound size, reduce the odor, lessen the discharge, and reduce the infection rate, but silver also improved the efficacy of a number of surgical procedures, including grafts, incisions, amputations, and prosthesis use. However, these many benefits of silver quickly faded from knowledge and use, following the introduction of antibiotics post-WW2. Fortunately, new research is bringing them back.

In fact, recent studies reveal that silver nanoparticles play a very key role in tissue regeneration and wound repair. Scientifically speaking, the silver particles help to significantly reduce what is known as the 'inflammatory cytokine cascade' following physical wounds. They decrease the inflammation process and help to facilitate quick, healthy, and effective tissue repair. Thus, your cuts and wounds heal faster, feel better, and scar less, because of silver nanoparticles.

But you don't have to be a wounded warrior or even scraped up to reap the restorative benefits of colloidal silver. Take, for instance, the anti-inflammatory powers of silver in general. Numerous researchers, peer-reviewed empirical studies, professional papers, literature reviews, and meta-analyses have found silver to be highly conducive to inflammation reduction.

So why is this important? Because inflammation is *everywhere*! The human body can become inflamed for a number of reasons. Biochemically speaking, inflammation is the result of white blood cells releasing substances to combat foreign organisms. However, these 'foreign organisms' take many forms that we don't consider. And sometimes, due to years of unhealthy living, our bodies inflame when they shouldn't!

Simply consider the modern diet. You're running around, busy at work, busy with family, busy with friends, doing this, that and the other and running out of time. Sometimes you're tired, so you grab a coffee, you eat a quick meal, you drink soda, you drink alcohol, you don't exercise when you could or sleep like you should. You cut corners and risk health, but you do it anyway (we all do) because heck, life gets crazy sometimes.

Today's diet filled with artificial ingredients, preservatives and additives is a main reason for inflammation. You may feel bloated. You may feel groggy. Your thoughts might be muddled and your mental and physical stamina depleted.

And even when you do start doing things right, such as exercising more and eating better, you still get inflamed. With exercise especially. When we exercise, the blood flows to the area of damage, and with proper recovery, repairs that area to be stronger. This is the basic science of building muscle, endurance, strength, cardio, etc. You first damage yourself, your body responds using vital vitamins, minerals and nutrients, and you recover to be stronger.

But sometimes recovery isn't what it should be. Sometimes we lag, or we do more harm than good, never giving our bodies and minds the 'fuel' they need to improve and sustain. Add in the fact that many of us have been *conditioned* to regard physical and mental stress the wrong way, and you've got yourself a recipe for disaster!

So use silver. Silver nanoparticles help with inflammation and tissue repair. They will expedite the healing process, reducing redness, stiffness, soreness and swelling. Whether applying colloidal silver topically to a specific area (sore arms, open cut, healing wound, etc.) or under the tongue to circulate it throughout, be mindful of your body.

If you're involved in a very physically laborious job, or you're an exercise junkie, or you're exposed to environmental toxins and chemicals, or you're just dealing with acute and chronic physical discomfort and/or pain – use silver!

More likely than not, it'll work like… 'Magic.'

(D) Infections of the ear – Remember, colloidal silver surpasses antibiotics because unlike antibiotics, silver is a broad spectrum antimicrobial and antifungal. That means it treats bacteria, viruses and fungi! Nowadays, ear infections are usually treated the same. Physicians prescribe an anti-inflammatory or painkiller, before even treating the infection site. This can lead to stomach and throat pain, and even cause toxicity in the liver. After this, physicians prescribe antibiotics, which truthfully may or may not target the specific pathogen. Sometimes, addition antibiotics get prescribed when the bacteria mutates, or the previous antibiotic does not adequately target it.

So then what? Well, now you've got a body (and mind) pumped full of pharmaceuticals, unnecessarily, and there is a growing chance that the mutated microbes will only intensify.

In the case of chronic ear infection where the fluid builds up behind the eardrum and presents a danger of bursting it, the doctor may prescribe on-going treatment with antibiotics. When these treatments fail, they may recommend surgery to remove adenoids and then you actually have to get your eardrum punctured for a drainage tube. In the end, it's not pretty and you run the risk of hearing loss – which was the darn thing you were trying to prevent in the first place!

So stop. Forget all this mess, and go the colloidal silver route. Tilt your head, administer 5 to 10 drops in the affected ear(s), let it work its way in for a couple minutes, then tip out. Do this several times a day until the symptoms disappear.

(E) UTI – Nobody wants to talk about this embarrassing condition, but when it happens, it hurts! So treat the burn.

Don't let a urinary tract infection keep you out of commission.

Instead of guzzling cranberry juice, or going on long-term drugs, try silver. Now, your first thought might be: but how? And no, the answer isn't what you think.

Although applying silver topically to your genitals might combat the microbes on the surface, urinary tract infections are primarily *inside* you.

So here's what you do: Spray or spoon in half a teaspoon of colloidal silver under your tongue. This ensures that it goes straight to your bloodstream and is not slowed or lost in less efficient delivery mechanisms. Once swallowed, wait however many hours until you notice more symptoms of the UTI. This may mean taking up to 2 teaspoons per day.

Also be mindful. Continue to force water and other liquids to clear our your urinary tract. Force yourself to drink and urinate as much as possible, and if so inclined, put half a teaspoon of colloidal silver into your water bottle as well.

Proceed daily for no more than two weeks and seek medical advice if needed.

(F) Kill the Flu! - That's right, sound too good to be true? Well believe it. Extensive studies indicate that silver actually has the ability to affect the protein structures of given viral surfaces, thus essentially 'turning off' the virus. This incredible finding is simply more evidence that high quality colloidal silver is a miraculous antimicrobial. If you're looking to kick the influenza in the butt, administer multiple drops in your mouth a couple times a day, and swallow. Maintain this regimen for no more than 10 days, and weigh the results before seeking further medical advice.

(G) Regular Immune Maintenance – Alright, so by now you realize that colloidal silver is integral to treating a variety of health and body conditions. However, what you may not realize is: you don't need 'conditions.' Don't save colloidal silver for when you *think* you need it – you *always* need it!

Colloidal silver should become part of your daily health and wellness regimen. It has numerous subtle and significant benefits for daily living. In fact, the immunomodulatory effects of colloidal silver are well-documented.

So don't delay. The trick is, to keep the dose low, only a couple drops a day, alternating days off and on, to sustain an optimal baseline. Not only will this keep your body's natural defense mechanisms and restorative processes running smoothly, it will also improve your mind as a result.

Just think about your average day. Your immune system is constantly under attack, fluctuating as the internal and external environments change. Chemicals, toxins, byproducts, stressors and free radicals impact us on a continuous, molecular basis. It isn't just how you feel or how your body functions, it's how you think. It's how you interact with your world, with coworkers and friends, loved ones and family. A weakened immune system will affect your feelings as much as your thoughts, your intelligence as much as your *emotional* intelligence.

Frankly, much of what is going on in your body you have no idea about. Sure, you'll notice more significant changes, whether good or bad, but it is the accumulative effects of negative actors that catch us off guard.

So don't be blindsided. Make small regular doses part of your regimen. And if you're worried you might be taking too much or are going in blind and clueless, make a simple calculation: multiply your body weight by 12, then divide by the total ppm of your colloidal silver product. The resulting number is the number of eye dropper drops you can take on a *daily* basis, non-stop, for your entire life. Of course, some people take many more without negative consequence.

Just be sure to check the label on your dropper, as some products use much greater dropper volumes. The typical eye dropper drop, however, contains 0.05 ml per drop. A typical teaspoon contains 5 ml. So in order to ingest 1 teaspoon of colloidal silver, you would need to to take 100 drops of the standard eye dropper.

 Now for some more math. If you're the average adult, you weigh ~ 140 lbs. So that's 140 x 12 = 1680. Now, let's say you're using a high quality colloidal silver with a concentration of 30 ppm – so that's 1680/30 = 56. In other words, you can take roughly 60 eye drops (at 0.05 ml per drop) per day, *every single day*, for your entire life, without any worries. That equals out to about 3 ml a day, or about half a teaspoon of colloidal silver per day, for life.

And if you're still not convinced, worried you might somehow overdose (virtually impossible), consider this: water, water, water. Drink plenty of water to flush any excess silver from your system. And don't forget to employ intermittent hydration cleanses to detox and energize your faculties. And if that isn't enough, consider even a <u>strong neurogenic diet</u> for maximized results.

Remember: If at any time you want to restrict your silver intake, free to cut back, or take extended breaks for days, weeks or months. Unlike many drugs, stopping the intake of silver abruptly will not have deleterious effects on your body or mind.

Overall, you should be fine, without any noticeable problems, and with plenty of perks!

(H) Stomach & Gastrointestinal – Nothing is worse than a bad stomach bug. You're curled over the toilet, praying to anything in sight that the suffering will stop. In many cases, you're totally dehydrated. Sometimes you continue to retch even though it seems every last bit of whatever was inside you, has left.

And even if stomach or gastrointestinal problems do not induce vomiting, they're still awful. You feel pain, you feel soreness, you fill discomfort that's hard to describe.

But don't worry. Silver is here to help. And not only does silver help with a number of common stomach bugs, stomach problems, and gastrointestinal issues, but it also fights some very strong, and dangerous strains.

For one, research indicates that nanosilver can actually inactivate drug-resistant forms of the highly dangerous E. Coli. This form of food poisoning can strike thousands of people with severe problems, becoming even lethal in some cases. Fortunately, silver appears to target it in ways many pharmaceuticals do not.

Other research has shown that silver particles can also drastically reduce the effects of salmonella. Colloidal silver has even been linked to the treatment of Crohn's disease, an inflammatory bowel disease occurring anywhere from the mouth to the rectum. Given what we already know about colloidal silver's antimicrobial and anti-inflammatory properties, experts contend that silver is also conducive in alleviating the abdominal distress, diarrhea, fever and chronic inflammation associated with Crohn's.

But that's not all! Silver nanoparticles are also extremely efficacious in treating *H. pylori*, which is a primary cause of chronic mucosal inflammation in the stomach and intestine, ulcers, and numerous other gastrointestinal abnormalities. So if you're struggling with all sorts of stomach or gut problems, whether it be serious microbial or fungal infestations, transient stomach issues, food poisoning, stomach flu, or chronic 'disorders' that physicians have told you will require long-term maintenance drugs – try silver! Simply use it as described in previous sections, administering 5 to 10 eye droppers sublingually several times a day for as many days as symptoms persist.

(I) Oral health – Swishing nanosilver, gargling nanosilver and brushing your teeth with nanosilver are all highly effective ways to improve dental hygiene and optimize oral health. In fact, nanosilver flouride particles have recently been applied for their antimicrobial, bioadhesive properties, in addition to their ability to slow and prevent dental decay without staining the teeth. And if you're opposed to flouride toothpastes and mouthwashes, you can readily find alternatives that contain nanosilver and iodine combinations.

Think of the long-term. You'll reduce plaque buildup, restore tender or bleeding gums, delay or avoid periodontal disease, and keep your teeth from decaying. Beyond your health and well-being, this will save you a lot of $$ in dental bills!

Nanosilver is also critical to dentistry in general. Whether you are getting a routine cleaning, having teeth pulled, getting braces, dentures, implants or some other oral intervention by endodontists and periodontists, nanosilver is critical to dental procedures.

So enjoy it. Ditch your normal toothpaste and mouthwash for a nanosilver alternative. Your mouth will feel fresher and your teeth will be cleaner, and your overall health will be better. Don't forget, oral health is absolutely indispensable to overall health, and can be the source of many problems in other, seemingly unrelated areas of the body.

So be smart! Go silver.

(J) Sores, boils, burns and rashes – We've already covered these effects to an extent, but just in case you weren't convinced… even more and more researchers today are finding out how colloidal silver can readily soothe and repair tissues and inflammation. For sunburn or other burns, use a dressing of silver. And if you don't feel like purchasing the proprietary dressing, simply squirt a little colloidal on your band aid, or directly on the affected area. Feel free to use this topical approach for sores, boils and rashes as well. Massage the silver and reapply 5-10 eye droppers several times a day.

And while you're nursing your wounded tissue, be sure to restrict exposure to strong sunlight. Instead, opt for intermittent exposure by optimizing Vitamin D3 absorption.

And let the silver work its magic!

(K) Fungal Infections – Are you dealing with athlete's foot, nail fungus, or some kind of other unsightly, unseemly, all-too-smelly fungal infection? Give silver a try. Again, apply droppers to the area repeatedly until the symptoms subside. Studies show that pathogenic fungus is heavily inactivated by silver ions.

So enjoy the remedy. Erase those nasty odors and watch those pesky infections fade away!

(L) Staph infections (MRSA) – In case you're unfamiliar with staph infections, consider yourself lucky. Although many people are carriers of staph bacteria, many of them will never experience a staph infection. Staph infections are caused by bacteria typically found on the skin and nose of even the healthiest individuals, and typically result in nothing more than minor skin irritation. However, when staph infections occasionally find their way into the body, they can affect the entire body, ravaging organs and potentially killing the carrier.

In most cases, antibiotics are used in conjunction with local draining to remove and kill the bacteria. However, this is not always effective. Sometimes the symptoms will persist as the staph bacteria mutate and become medicine-resistant. In these cases, a plethora of symptoms will exist. Common symptoms are skin-related, such as boils, painful rashes, swelling and redness, blisters, and oozing sores accompanied by fever.

Unfortunately, it gets worse than that. Staph bacteria have also been linked to food poisoning, blood poisoning (affecting vital organs such as your heart and brain), life-threatening shock syndrome causing sudden vomiting, fever and confusion, and even severe arthritic joint swelling. And MRSA, a type of staph bacteria highly resistant, can bring about all these things.

That's why silver is crucial. Clinical studies reveal that nanoparticles of colloidal silver literally eradicate MRSA, even in its most multiresistant forms. The findings, truly, are incredible.

(M) Yeast Infections – Continuing on infections, colloidal silver is also great for yeast. The antimicrobial, antifungal properties of high quality colloidal silver are truly remarkable when it comes to battling, weakening and even inactivating yeast in your gut. Gut flora, and the general health of your gastrointestinal system, is critical for overall well-being. Gut health is even linked directly to your brain, and indirectly, the quality and effectiveness of your thoughts, moods, and perceptions. One of the main yeasts that feed on your gut is Candida albicans. It is one of the most powerful and prevalent, and is quickly destroyed by silver nanoparticles. In fact, silver particles are potent inhibitors of C. albicans biofilm formation.

So whether you have yeast infection in the gut, or yeast infection elsewhere, colloidal silver is of likely benefit. Apply it topically if the infection is visible and on the surface. Ingest silver if your infection is in your body and organs.

(N) Herpes – That's right, high quality colloidal silver may even treat STDs. Don't believe it? Well just consider the fact that a related compound, silver nitrate, has been shown to inactivate the Herpes simplex virus.

(O) Cancer – Now this can't be true, can it? Well, according to a number of researchers, silver complexes have consistently been shown to strengthen cancer treatments such as chemotherapy, while also working to destroy cancer cells on their own right. The research is still young and burgeoning, but if it can be shown that silver compounds and colloids significantly affect cancer cells, the medical community might be in for a revolution!

(P) Difficulty breathing – Whether allergies, toxic reaction, or asthma, trouble breathing is annoying, scary, and potentially deadly. But fortunately, ingesting silver nanoparticles appears to modulate pre-existing medication, and also offer strong standalone effects on asthma and asthma-like symptoms.

(Q) Hair Loss – We all hate looking in the mirror and seeing our hair thinning. The lost of luster, the receding hairline, the patchy baldness, the showing of scalp, that leery feeling as age finally, truly, settles in. But what if there were a way to slow it, without combing through the thousands of supposed worthwhile shampoos, conditioners and DIY remedies? Simply spray on the problem areas 3 to 4 times a day, daily, as part of your naturopathic hair remedy regimen, and monitor for the next two weeks. You might just notice new hairs sprouting up!

(R) Household use – So far we've been talking about how silver has tremendous benefits for human health. But let us not forget, that human health is impacted by factors well outside the human body. In many cases, the place you stay, the place you call home, may be killing you. Don't like hearing it? Well neither does anybody, but that doesn't make it false. Truth be told, your house, your apartment, your condo or duplex is constantly under siege from a number of environmental forces. And when those forces take over…

Just think about all the structural problems that can go wrong. You can have plumbing leaks, bathroom leaks, roof problems, air leak problems, bad ventilation and problems with your walling.

When these structural problems become too much, your place of residence becomes a hotbed for toxins. You get spores, you get mold, and in the worst cases, you get potentially deadly black mold. In case you didn't know, black mold loves humid dark places. It gets in your air without you even knowing, and then suddenly you're feeling sick more often, you're coughing, you have a runny nose, you have trouble sleeping, concentrating, and getting yourself going. Your thoughts changes, your emotions change, your moods change. Black mold has not only been linked to a number of common and degenerative health conditions, but even many neurological problems, such as ADHD in children!

So don't succumb to black mold. Don't let it proliferate without your knowing. If you don't have the money for expensive repairs or remodeling, use silver. To remove the mold and mildew, spray some shots of top quality colloidal silver on the area, let it sit for 15 minutes, then wipe away.

And that's all.

(S) Preservation – One final area of silver use, outside of the human body, is for preserving condiments and foods. For centuries, the elites and royals have hogged silver for themselves. They've eaten with silver spoons, forks, knives and chopsticks. They've sipped form silver goblets. They've stored food, water and wine in silver vessels – and they've relished every second.

And so can you. If you're on a budget, ditch the expensive doomsday preparedness equipment and store with silver. Preserve your favorite foods, drinks and condiments simply:

For foods, use 1 teaspoon of 10 ppm colloidal silver for every quart if canning. Then seal the containers as you would normally. For liquids, use 1 oz of 10 ppm colloidal silver for every gallon, mix it well, and seal the containers as you would normally. You can also add smaller amounts of colloidal silver to products in your refrigerator or cabinets that are *not* being stored long-term. Because of colloidal silver's remarkable antimicrobial and antifungal properties, the shelf-life of many of your foods will be extended significantly.

You'll be surprised!

But that's what it's all about, isn't it?

***

A Future With a Silver-lining

It can't be called magic if it doesn't catch you off guard. If it doesn't, at least a little bit, leave us wondering… How, does, it, do it?

Truth is, colloidal silver and silver nanoparticles are a miracle of modern nature and science. We are simply taking natural properties gifted to us by the world around us, and using them for their intended purpose. For centuries, beyond thousands of years, this amazing metal has imbued its healing, resonating, all-encompassing powers in ways we could never truly fathom. It has lengthened lives, and prevented deaths. It has restored hope to the suffering, to the cynics, to the naysayers and know-nothings. It's brought energy to the waking world, as it's eased others into the sleeping world. Silver, in its innumerable forms and fashions, is a harbinger of things good.

So don't believe the lies. Don't fall victim like a mindless automaton, convinced that silver is your enemy, nothing more than a hyped-up line of 'quack' designed to empty your wallet and dampen your mind.

The Powers-That-Be have long tried to squash the truth.

But that truth is shining bright. Now, more than ever, cutting-edge research is bringing that truth to light. Innovative biotech companies, manufacturers, nutritionists, physicians, specialists, and naturopaths are coming together, working intently in concert to create a culture of acceptance.

The time has come to understand, accept and celebrate the good that silver can bring. It is time to integrate our collective minds, and efforts, and bodies, as one, in order to organically and safely incorporate the colloid.

We are all in this thing called life together. And life is crazy. And although silver may not treat every condition and problem in life, and although it might work okay for some and amazingly for others, this is no reason to slow. We must expedite the process of research and implementation, bringing silver to the fore.

In the end, colloidal silver is an amazing gift. It is truly a *game-changer*. I personally believe that it can revamp and transform the lives of many of you, mine included, and it is for this sole, critical reason, that I have written this book.

But please, do not trust me alone. Do your own research, understand what you are learning, and understand what you have learned. Take heed form the experts, scour the resources, and experiment safely and progressively. If you incorporate colloidal silver in the ways I have enumerated thus far, I have every confidence that you will improve your life immeasurably.

This life is yours to lead. Make it Magical.

A Special Note:

Thank you for reading *"Silver Magic: How Colloidal Silver Can TRANSFORM Your Life"* If you enjoyed reading this book and would like to be included on an email list for when similar content is available, feel free:

SUBSCRIBE

As always, thank you for reading. And may you continue to live healthily and happily.

Sincerely,

C.K. Murray

Other works by C.K. Murray:

1. *Mindfulness Explained: The Mindful Solution to Stress, Depression, and Chronic Unhappiness*

2. *Emotional Intelligence Explained: How to Master Emotional Intelligence and Unlock Your True Ability*

39. Body Language Explained: How to Master the Power of the Unconscious

www.ingramcontent.com/pod-product-compliance
Lightning Source LLC
Chambersburg PA
CBHW071229220526
45468CB00002B/776